Biological Life And Electricity

By Ian Beardsley

ISBN: 978-1-67818-734-7

ENTITY BOOKS

TABLE OF CONTENTS

BONE

Equation of Density and Molar Mass................5

The Intermembral Index...............................11

Appendix..22

SILICON AND CARBON

Silicon and Carbon....................................37

Amino Acids...49

DNA...58

Male and Female in AI.................................63

ASYMMETRY IN AI

Artificial Intelligence...................................68

Biological Life...78

Differential Equations.................................82

Cross Product...86

Integrating..89

BONE

EQUATION OF DENSITY AND MOLAR MASS

Density of silicon is Si=2.33 grams per cubic centimeter.

Density of germanium is Ge=5.323 grams per cubic centimeter.

Density of hydroxyapatite is HA=3.00 grams per cubic centimeter.

This is

$$\frac{3}{4}Si + \frac{1}{4}Ge \approx HA \quad \text{where} \quad HA = Ca_5(PO_4)_3OH$$

Where HA is the mineral component of bone, Si is an AI semiconductor material and Ge is an AI semiconductor material. This means

$$\frac{Si}{HA}Si + \left[1 - \frac{Si}{HA}\right]Ge = HA$$

The harmonic mean between Si and Ge is HA,...

$$\frac{2SiGe}{Si + Ge} \approx HA$$

While

$Si \approx Ge - HA$ and,

$HA \approx \dfrac{2SiGe}{Si + Ge}$ yield

$Ge^2 - 2SiGe - Si^2 = 0$ With solution

$\dfrac{Si}{Ge} \approx \dfrac{1}{\sqrt{2} + 1}$ We have

$Si \approx Ge - HA$ and,

$HA \approx \dfrac{2SiGe}{Si + Ge}$ and,

$\dfrac{3}{4}Si + \dfrac{1}{4}Ge = HA$ yield

$Si = \dfrac{2}{5}Ge$ And we have

$\dfrac{1}{\sqrt{2} + 1} = 0.414$

$\dfrac{2}{5} = 0.4$ Which are close to the same by 96.6%:

$\dfrac{0.4}{0.414} = 0.966$

And, the equations work approximately for both density and molar mass:

$$2.33 g/cm^3 = \frac{2}{5} 5.323 g/cm^3 = 2.129 g/cm^3$$

$$\frac{2.33}{2.129} = 0.91 \quad \text{(91\% accuracy)}$$

$$28.09 g/mol = \frac{2}{5} 72.61 g/mol = 29.044 g/mol$$

$$\frac{28.09}{29.044} = 0.967 \quad \text{(96.7\% accuracy)}$$

$$\frac{3}{4} Si + \frac{1}{4} Ge \approx HA$$

Implies

$$\frac{Si}{HA} Si + \left[1 - \frac{Si}{HA} \right] Ge = HA$$

And this yields the sextic:

$$Si^2(Si + Ge)^4 - SiGe(Si + Ge)^4 + 2SiGe^2(Si + Ge)^3 - 4Si^2Ge^2(Si + Ge)^2 = 0$$

For which

$$\frac{Si}{Ge} = \frac{1}{1 - \sqrt{2}}$$

$$\frac{Si}{Ge} = \frac{1}{1 + \sqrt{2}}$$

Si=-Ge

Si=Ge

Si=0, Ge=0

3D plot:

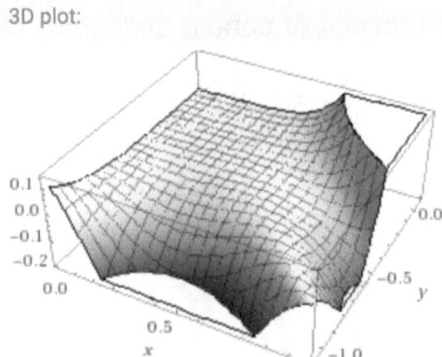

Contour plot:

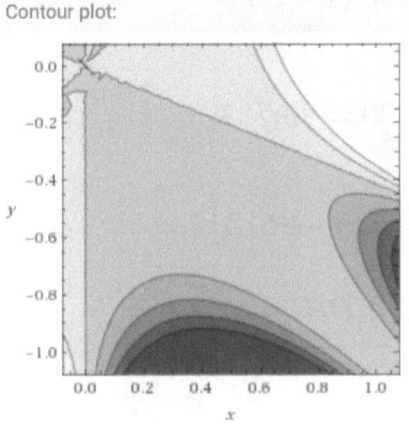

The mineral component of bone hydroxyapatite (HA) is

$$Ca_5(PO_4)_3OH = 502.32\frac{g}{mol}$$

The organic component of bone is collagen which is

$$C_{57}H_{91}N_{19}O_{16} = 1298.67\frac{g}{mol}$$

We have

$$\frac{Ca_5(PO_4)_3OH}{C_{57}H_{91}N_{19}O_{16}} = 0.386795722$$

$$\phi = 0.618033989$$

$$1 - \phi = 0.381966011$$

$$\frac{Ca_5(PO_4)_3OH}{C_{57}H_{91}N_{19}O_{16}} \approx (1 - \phi)$$

$$\frac{0.381966011}{0.386795722}100 = 98.75\%$$

$$\frac{Si}{Ge} = \frac{28.09}{72.61} = 0.386861314 \approx (1 - \phi)$$

$$\frac{Si}{Ge} \approx \frac{Ca_5(PO_4)_3OH}{C_{57}H_{91}N_{19}O_{16}}$$

THE INTERMEMBRAL INDEX

I believe it is best to compare biological life to another construct, like artificial intelligence (AI) That is, intermembral index to silicon (si) and germanium (ge).

Chimpanzee index is 106 or 1.06 in otherwords as a fraction, meaning their forelimbs are longer than their hind limbs compared to humans, which are around 68-70 or 0.68 to 0.70 meaning their hindlimbs are longer than their forlimbs. Thus we see how their forelimbs longer for climbing, arm hanging and swinging activities. The longer hindlimbs of humans mean they depend soley on these for propulsion in bipedal walking. Lucy, the 8.2 million year old hominid (Australopithecus Afarensis) has index 88 intermediate between humans and chimpanzees (0.88) and this due to a shortened humerous, not elongated thigh, showing arm length reduced first in the evolutionary trend toward being bipedal. She probably used hindlimbs for bipedal propulsion and forelimbs for climbing.

CHIMPANZEE

HUMAN

22"

32"

humerous + radius = 22" and femur + tibia = 32" intermembral index $(i) = 22/32 = 0.6875$

$$\frac{1}{i} \approx \frac{1}{\sqrt{2}+1} + 1 = \sqrt{2} \qquad \frac{si}{ge} \approx \frac{1}{\sqrt{2}+1}$$

LUCY

The intermembral index compares the forelimbs of vertebrates to their hindlimbs. A ratio greater than one means the forelimbs are longer than the hindlimbs and less than one the hindlimbs are longer. It is this ratio that tells paleontologists a great deal about the manner of propulsion of a vertebrate.

$$Si \approx Ge - HA$$

$$HA \approx \frac{2SiGe}{Si + Ge}$$

$$\frac{Ge^2 - 2SiGe - Si^2}{Si + Ge} = 0$$

$$x^2 - 2xy - y^2 = 0$$

$$\frac{y}{x} = \frac{1}{\sqrt{2} + 1}$$

Since the intermembral index, i, for humans is

$$i \approx 0.7$$

And,

$$\frac{Si}{Ge} = \frac{1}{\sqrt{2} + 1}$$

By both density and molar mass, we have

$$\frac{1}{i} \approx \frac{1}{\sqrt{2} + 1} + 1$$

Now we introduce the organic component of bone, collagen:

$$Ca_5(PO_4)_3OH = 502.32\frac{g}{mol}$$

$$C_{57}H_{91}N_{19}O_{16} = 1298.67\frac{g}{mol}$$

We have

$$\frac{Ca_5(PO_4)_3OH}{C_{57}H_{91}N_{19}O_{16}} = 0.386795722$$

$$\phi = 0.618033989$$

$$1 - \phi = 0.381966011$$

$$\frac{Ca_5(PO_4)_3OH}{C_{57}H_{91}N_{19}O_{16}} \approx (1 - \phi)$$

$$\frac{0.381966011}{0.386795722}100 = 98.75\%$$

$$\frac{Si}{Ge} = \frac{28.09}{72.61} = 0.386861314 \approx (1 - \phi)$$

$$\frac{Si}{Ge} \approx \frac{Ca_5(PO_4)_3OH}{C_{57}H_{91}N_{19}O_{16}}$$

By molar mass. Where, ϕ is the golden ratio conjugate, is recurrent throughout ratios in vertebrates determined by bone lengths.

We have said

$$\frac{1}{i} \approx \frac{1}{\sqrt{2}+1} + 1 \approx \sqrt{2}$$

$$\frac{Si_\rho}{Ge_\rho} \approx \frac{1}{\sqrt{2}+1}$$

So (we have subscripted Si and Ge with rho for density),...

$$\frac{1}{i} \approx \frac{Si_\rho}{Ge_\rho} + 1 \quad \text{Or...}$$

$$\frac{1}{i} \approx \frac{Si_\rho + Ge_\rho}{Ge_\rho}$$

$$i \approx \frac{Ge_\rho}{Si_\rho + Ge_\rho}$$

We have also said (subscripting with M for molar mass)

$$\frac{Si_M}{Ge_M} \approx \frac{Ca_5(PO_4)_3OH}{C_{57}H_{91}N_{19}O_{16}}$$

$$\frac{Ge_\rho}{Si_\rho + Ge_\rho} \approx \frac{\sqrt{2}}{2} = sin\frac{\pi}{4} = cos\frac{\pi}{4}$$

$$\frac{Si_M}{Ge_M} \approx (1 - \phi)$$

$$Si_M \approx (1 - \phi)Ge_M$$

$$Si_M \approx \frac{Ca_5(PO_4)_3OH}{C_{57}H_{91}N_{19}O_{16}}Ge_M$$

In order for Si and Ge to semiconductor they must be doped. Often the doping agents are phosphorus (P) and boron (B). We find by atomic radius where SI_R=110pm, Ge_R=125pm, P_R=100pm and B_r=85pm

$$\frac{P_R + B_R}{Si_R} \approx \Phi \quad or,.... \quad \frac{Si_R}{P_R + B_R} \approx \phi$$

But if we take the arithmetic mean between the geometric mean of P_M and B_M and the harmonic mean and divide by Si_M we find that,...

$$\frac{\sqrt{P_M B_M}(P_M + B_M) + 2P_M B_M}{2(P_M + B_M)Si_M} \approx \phi$$

Which yields,...

$$\frac{2Si_R}{P_R + B_R}Si_M \approx \sqrt{P_M P_M} + \frac{2P_M B_M}{P_M + B_M}$$

Thus,...

$$\frac{2Si_R}{P_R + B_R} \approx \frac{C_{57}H_{91}N_{19}O_{16}}{Ca_5(PO_4)_3OH}\left[\sqrt{P_M B_M} + \frac{2P_M B_M}{P_M + B_M}\right]\frac{1}{Ge_M}$$

$$\frac{Ge_\rho}{Si_\rho + Ge_\rho} \approx \frac{\sqrt{2}}{2}$$

Thus since we have

$$\frac{2Si_R}{P_R + B_R} \approx \frac{C_{57}H_{91}N_{19}O_{16}}{Ca_5(PO_4)_3OH}\left[\sqrt{P_M B_M} + \frac{2P_M B_M}{P_M + B_M}\right]\frac{1}{Ge_M}$$

$$\frac{Ge_\rho}{Si_\rho + Ge_\rho} \approx \frac{\sqrt{2}}{2}$$

And, since we have,...

$$CO_2 = 12.01 + 32.00 = 44.01$$

$$O_2 = 32.00$$

$$\frac{CO_2}{O_2} = \frac{44.01}{32.00} = 1.375 \approx 1.414 = \sqrt{2}$$

Then we can write as well

$$\frac{2Si_R}{P_R + B_R} \approx \frac{C_{57}H_{91}N_{19}O_{16}}{Ca_5(PO_4)_3OH}\left[\sqrt{P_M B_M} + \frac{2P_M B_M}{P_M + B_M}\right]\frac{1}{Ge_M}$$

$$\frac{2Ge_\rho}{Si_\rho + Ge_\rho} \approx \frac{CO_2}{O_2}$$

CO2 (carbon dioxide) is what animal life exhales when it breathes, which is taken up by plant life to make the oxygen gas (O2) that it breathes in a process called photosynthesis where plants get energy from the sun to make the monomer used to make its food the carbohydrate CH2O at the basis of making the more complex sugars at the bottom of the food chain. The reaction is:

$$CO_2 + 2H_2O + photons \longrightarrow CH_2O + O_2 + H_2O$$

we should consider the densities of
Si and Ge, which are written Siρ and
Geρ, and the molar masses of Si and Ge,
which are written Si$_M$ and Ge$_M$, with
respect to one another:

$$\frac{Si\rho}{Ge\rho} \approx \frac{1}{\sqrt{2}+1}$$

$$\frac{Si_M}{Ge_M} \approx 1-\phi$$

where $\phi = \dfrac{\sqrt{5}-1}{2}$

and $\dfrac{\sqrt{2}}{2} = \sin 45° = \cos 45°$

$\sqrt{2}$ is the ration of the diagonal
of a square to its sides and is
two right triangles with angles 45°
each:

And ϕ is
formed by
the triangle
with legs 1
and 2:

The intermembral index, i, fo
humans is $i = \dfrac{\sqrt{3}}{2} = 0.7$

$$\frac{Si_\beta}{Ge_\rho} = \frac{1}{\sqrt{2}+1} \cdot \left[1 - \frac{\sqrt{5}-1}{2} \right] \quad \text{Interestingly}$$

$$2\cos 45° = 2\cos\frac{\pi}{4} = \sqrt{2}$$

$$2\cos 36° = 2\cos\frac{\pi}{5} = \Phi$$

where $\Phi = \frac{\sqrt{5}+1}{2} = \frac{1}{\phi}$

$$2\cos 30° = 2\cos\frac{\pi}{6} = \sqrt{3}$$

we can show

$$-2\cos\left(\frac{\pi}{4}\right) + 2\cos\left(\frac{\pi}{5}\right) + 2\cos\left(\frac{\pi}{6}\right) \approx \frac{air}{H_2O}$$

by molar mass where $air = 0.25 O_2 + 0.75 N_2$

$$= 29.0 \text{ g/mol}$$

Thus where $\frac{\sqrt{2}}{2}$ is i in humans

$\frac{\sqrt{5}-1}{2}$ is in humans as well where

the distance from head to navel compared to navel to bottom of the feet is ϕ.

$i = \frac{a}{b}$

$\phi = \frac{c}{d}$

Thus we have the very interesting equations

$$\left[1 - \frac{Si_\rho}{Ge_\rho}\right] \approx \frac{(1 - \phi)}{i}$$

$$i = \cos\frac{\pi}{4}, \phi = 2\cos\frac{\pi}{5} - 1$$

$$\frac{Si_\rho}{Ge_\rho}\frac{Si_M}{Ge_M} \approx \frac{(1 - \phi)}{(2i + 1)}$$

APPENDIX

$$\frac{Si}{HA}Si + \left[1 - \frac{Si}{HA}\right]Ge = HA$$

$$\frac{Si^2}{HA} + Ge - \frac{Si}{HA}Ge \approx HA$$

$$\frac{1}{HA}Si^2 - \frac{Ge}{HA}Si + Ge \approx HA$$

$$\frac{1}{HA^2}Si^2 - \frac{Ge}{HA^2}Si + \frac{Ge}{HA} \approx 1$$

$$\frac{1}{HA^2}Si^2 - \frac{Ge}{HA^2}Si + \frac{Ge}{HA} - 1 \approx 0$$

$$\frac{1}{HA^2}Si^2 - \frac{Ge}{HA^2}Si + \left[\frac{Ge}{HA} - 1\right] = 0$$

$$(x + a)(x + a) = x^2 + 2ax + a^2$$

$$(x + a)^2 = x^2 + 2ax + a^2$$

We see that the square of the binomial is a quadratic where the third term is the square of one half the middle coefficient. This gives us a method to solve quadratics called completing the square:

$$ax^2 + bx + c = 0$$

$$ax^2 + bx = -c$$

$$x^2 + \frac{b}{a}x = -\frac{c}{a}$$

$$\left(\frac{1}{2}\frac{b}{a}\right)^2 = \frac{1}{4}\frac{b^2}{a^2}$$

$$x^2 + \frac{b}{a}x + \frac{1}{4}\frac{b^2}{a^2} = -\frac{c}{a} + \frac{1}{4}\frac{b^2}{a^2}$$

$$\left(x + \frac{1}{2}\frac{b}{a}\right)^2 = \frac{b^2 - 4ac}{4a^2}$$

$$x + \frac{b}{2a} = \pm\frac{\sqrt{b^2 - 4ac}}{2a}$$

$$x = \frac{-b \pm \sqrt{b^2 - 4ac}}{2a}$$

$$\frac{1}{HA^2}Si^2 - \frac{Ge}{HA^2}Si + \left[\frac{Ge}{HA} - 1\right] = 0$$

$$x = \frac{-b \pm \sqrt{b^2 - 4ac}}{2a}$$

$$a = \frac{a}{HA^2} \qquad b = -\frac{Ge}{HA^2} \qquad c = \left[\frac{Ge}{HA} - 1\right]$$

$$b^2 - 4ac = \frac{Ge^2}{HA^4} - 4\frac{1}{HA^2}\left[\frac{Ge}{HA} - 1\right]$$

$$= \frac{Ge^2}{HA^4} - \frac{4Ge}{HA^3} + \frac{4}{HA^2}$$

$$= \frac{1}{HA^2}\left[\frac{Ge^2}{HA^2} - \frac{4Ge}{HA} + 4\right]$$

$$\sqrt{b^2 - 4ac} = \frac{1}{HA}\sqrt{\left(\frac{Ge}{HA} - 2\right)^2}$$

$$x = \frac{\frac{Ge}{HA^2} \pm \frac{1}{HA}\left[\frac{Ge}{HA} - 2\right]}{\frac{2}{HA^2}}$$

$$= \frac{1}{2}Ge \pm \frac{1}{2}HA\left[\frac{Ge}{HA} - 2\right]$$

$$= \frac{1}{2}Ge \pm \frac{1}{2}Ge - HA$$

$$Si = \frac{1}{2}Ge + \frac{1}{2}Ge - HA$$

$$Si = Ge - HA$$

$$Si \approx Ge - HA$$

$$HA \approx \frac{2SiGe}{Si + Ge}$$

$$Si \approx Ge - \frac{2SiGe}{Si + Ge}$$

$$\frac{(Si + Ge)Ge}{Si + Ge} - \frac{(Si + Ge)Si}{Si + Ge} - \frac{2SiGe}{Si + Ge} = 0$$

$$\frac{Ge^2 - 2SiGe - Si^2}{Si + Ge} = 0$$

$$x^2 - 2xy - y^2 = 0$$

$$x^2 - 2xy = y^2$$

$$x^2 - 2xy + y^2 = 2y^2$$

$$(x - y)^2 = 2y^2$$

$$x - y = \pm \sqrt{2}y$$

$$x = y + \sqrt{2}y$$

$$x = y(1 + \sqrt{2})$$

$$\frac{x}{y} = 1 + \sqrt{2}$$

$$\frac{y}{x} = \frac{1}{\sqrt{2} + 1}$$

$$\frac{Si}{Ge} \approx \frac{1}{\sqrt{2}+1}$$

A ratio is $\dfrac{a}{b}$ and a proportion is $\dfrac{a}{b} = \dfrac{b}{c}$ which means a is to b as b is to c.

The Golden Ratio (Φ)

$$\frac{a}{b} = \frac{b}{c} \quad \text{and.} \quad a = b + c$$

$$ac = b^2 \quad \text{or} \quad c = \frac{b^2}{a}$$

$$a = b + \frac{b^2}{a}$$

$$\frac{b^2}{a} - a + b = 0$$

$$\frac{b^2}{a^2} - 1 + \frac{b}{a} = 0$$

$$\left(\frac{b}{a}\right)^2 + \frac{b}{a} - 1 = 0$$

$$\left(\frac{b}{a}\right)^2 + \frac{b}{a} + \frac{1}{4} = 1 + \frac{1}{4}$$

$$\left(\frac{b}{a} + \frac{1}{2}\right)^2 = \frac{5}{4}$$

$$\frac{b}{a} = -\frac{1}{2} \pm \frac{\sqrt{5}}{2} \qquad \frac{b}{a} = \frac{\sqrt{5}-1}{2} \qquad \frac{a}{b} = \frac{\sqrt{5}+1}{2}$$

$$\phi = \frac{\sqrt{5} - 1}{2} \qquad \Phi = \frac{\sqrt{5} + 1}{2} \qquad \phi = \frac{1}{\Phi}$$

$$\frac{3}{4}Si + \frac{1}{4}Ge = HA$$

$$Si = Ge - HA$$

$$HA = \frac{2SiGe}{Si + Ge}$$

$$3Si + Ge = 4HA$$

$$Si + \frac{Ge}{3} = \frac{4}{3}HA$$

$$Si = \frac{4}{3}HA - \frac{Ge}{3}$$

$$Ge - HA = \frac{4}{3}HA - \frac{Ge}{3}$$

$$Ge + \frac{Ge}{3} = HA + \frac{4}{3}HA$$

$$\frac{3Ge + Ge}{3} = \frac{3HA + 4HA}{3}$$

$$4Ge = 7HA$$

$$Ge = \frac{7}{4}HA$$

$$Ge = \frac{7}{4}\frac{2SiGe}{Si + Ge}$$

$$Ge = \frac{14SiGe}{4Si + 4Ge}$$

$$1 = \frac{14Si}{4Si + 4Ge}$$

$$\frac{4}{14}(Si + Ge) = Si$$

$$4(Si + Ge) = 14Si$$
$$Si - \frac{4}{14}Si = \frac{4}{14}Ge$$

$$\frac{14Si - 4Si}{14} = \frac{4}{14}Ge$$

$$\frac{10}{14}Si = \frac{4}{14}Ge$$

$$5Si = 2Ge$$

$$\frac{Si}{Ge} = \frac{2}{5}$$

$$Si = \frac{2}{5}Ge$$

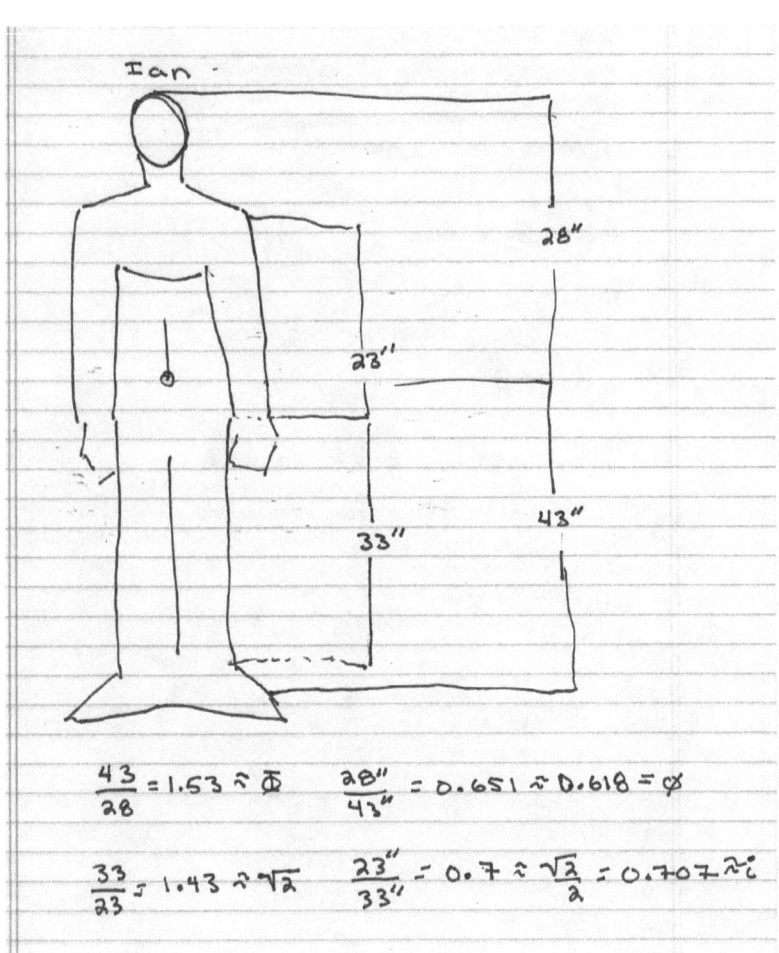

Ian -

$\frac{43}{28} = 1.53 \approx \Phi$ $\frac{28''}{43''} = 0.651 \approx 0.618 = \phi$

$\frac{33}{23} = 1.43 \approx \sqrt{2}$ $\frac{23''}{33''} = 0.7 \approx \frac{\sqrt{2}}{2} = 0.707 \approx i$

$a + b = 23 + 33 = 56 = A$

$c + d = 28 + 43 = 71 = B$

$\frac{A}{B} = 0.7887$ $\frac{B}{A} = 1.267857$

$a =$ humerous + radius

$b =$ Femur + tibia

$\frac{a+b}{c+d} = 0.7887 = \frac{56}{71}$

arm $=$ ~~24/22/23~~ $a =$ ~~24~~ 23 ~~23~~

leg $=$ ~~33mm~~ $b =$ ~~mm~~ 33 mm

humerous $= 13''$ $10 + 13 = 23$
radius $= 10''$

femur $= 17''$ $16 + 17 = 33$
tibia $= 16''$

humerous $= w$ $a = w + x$
radius $= x$

femur $= y$ $b = y + z$
tibia $= z$

Let's say here what is trying to
be expressed here is that

$$\frac{B}{A} = \frac{5}{4}$$

All of these ratios can be taken
as levers for which there are three
cases, each serving different ends.

1) The fulcrum is between the effort
   and the load for which the mechanical
   advantage may be greater than 1,
   less than 1, or equal to 1. A crow bar.

2) The load is between the effort
   and the fulcrum. The mechanical
   advantage is always greater than 1.
   It is a force multiplier. A wheelbarrow.

3) The effort is between the fulcrum
   and the load. It is a speed multiplier.
   A pair of tweezers, a hammer.

SILICON AND CARBON

SILICON AND CARBON

This is the prequel to *Bone*, where I compared the aspect of biological life that is bone, to artificial intelligence (AI). This work deals with the comparison of biological life to artificial intelligence where the elements with which the former are built (CHNOPS) to the elements with which the latter are built (Si, Ge, P, B, Ga, As) are concerned. It is a purpose of biological life (C, N, O, H) to discover the properties of (P, B, Si) so it can make computing machines which are necessary to its survival.

The golden ratio and the golden ratio conjugate are the solution of the quadratic

$$\left(\frac{a}{b}\right)^2 - \frac{a}{b} - 1 = 0 \text{ that meets the conditions}$$

$$\frac{a}{b} = \frac{b}{c} \text{ and } a=b+c$$

Where $\Phi = \frac{a}{b}$ and $\phi = \frac{\sqrt{5}-1}{2}$, $\phi = \frac{1}{\Phi}$.

We guess that artificial intelligence (AI) has the golden ratio, or its conjugate in its means geometric, harmonic, and arithmetic by molar mass by taking these means between doping agents phosphorus (P) and boron (B) divided by semiconductor material silicon (Si) :

$$\frac{\sqrt{PB}}{Si} = \frac{\sqrt{(30.97)(10.81)}}{28.09} = 0.65$$

$$\frac{2PB}{P+B}\frac{1}{Si} = \frac{2(30.97)(10.81)}{30.97+10.81}\frac{1}{28.09} = 0.57$$

$$\frac{0.65+0.57}{2} = 0.61 \approx \phi$$

Which can be written

$$\frac{\sqrt{PB}(P+B)+2PB}{2(P+B)Si} \approx \phi$$

We see that the biological elements, H, N, C, O compared to the AI elements P, B, Si is the golden ratio conjugate (phi) as well:

$$\frac{C + N + O + H}{P + B + Si} \approx \phi$$

So we can now establish the connection between artificial intelligence and biological life:

$$(P + B + Si)\frac{\sqrt{PB}(P + B) + 2PB}{2(P + B)Si} \approx (C + N + O + H)$$

Which can be written:

$$\sqrt{PB}\left[\frac{P}{Si} + \frac{B}{Si} + 1\right] + \frac{2PB}{P+B}\left[\frac{P}{Si} + \frac{B}{Si} + 1\right] \approx 2HCNO$$

Where HNCO is isocyanic acid, the most basic organic compound. We write in the arithmetic mean:

$$\left[\sqrt{PB} + \frac{2PB}{P+B} + \frac{P+B}{2}\right]\left[\frac{P}{Si} + \frac{B}{Si} + 1\right] \approx 3HNCO$$

Which is nice because we can write in the second first generation semiconductor as well (germanium) and the doping agents gallium (Ga) and arsenic (As):

$$\left[\sqrt{PB} + \frac{2PB}{P+B} + \frac{P+B}{2}\right]\left[\frac{P}{Si} + \frac{B}{Si} + 1\right] \approx HNCO\left[\frac{Ga}{Ge} + \frac{As}{Ge} + 1\right]$$

Where

$$\frac{Zn}{Se} \approx \frac{\left[\frac{P}{Si} + \frac{B}{Si} + 1\right]}{\left[\frac{Ga}{Ge} + \frac{As}{Ge} + 1\right]}$$

Where ZnSe is zinc selenide, an intrinsic semiconductor used in AI, meaning it doesn't require doping agents. We now have:

$$\sqrt{PB}\left(\frac{Zn}{Se}\right) + \frac{2PB}{P+B}\left(\frac{Zn}{Se}\right) + \frac{P+B}{2}\left(\frac{Zn}{Se}\right) \approx HNCO$$

We could begin with semiconductor germanium (Ge) and doping agents gallium (Ga) and phosphorus (P) and we get a similar equation:

$$\frac{2GaP}{Ga+P} = 42.866, \quad \sqrt{GaP} = 46.46749$$

In grams per mole. Then we compare these molar masses to the molar masses of the semiconductor material Ge:

$$\frac{2GaP}{Ga+P}\frac{1}{Ge} = \frac{42.866}{72.61} = 0.59$$

$$\sqrt{GaP}\frac{1}{Ge} = \frac{46.46749}{72.61} = 0.64$$

Then, take the arithmetic mean between these:

$$\frac{0.59+0.64}{2} = 0.615$$

We then notice this is about the golden ratio conjugate, ϕ, which is the inverse of the golden ratio, Φ. $\phi \approx \frac{1}{\Phi}$. Thus, we have

1. $$\frac{\sqrt{GaP}(Ga+P)+2GaP}{2(Ga+P)Ge} \approx \phi$$

2. $$\frac{\sqrt{GaP}(Ga+P)+2GaP}{2(Ga+P)Si} \approx \Phi$$

This is considering the elements of artificial intelligence (AI) Ga, P, Ge, Si. Since we want to find the connection of artificial intelligence to biological life, we compare these to the biological elements most abundant by mass carbon (C), hydrogen (H), nitrogen (N), oxygen (O), phosphorus (P), sulfur (S). We write these CHNOPS (C+H+N+O+P+S) and find:

$$\frac{CHNOPS}{Ga + As + Ge} \approx \frac{1}{2}$$

A similar thing can be done with germanium, Ge, and gallium, Ga, and arsenic, As, this time using CHNOPS the most abundant biological elements by mass:

$$\left[\sqrt{GaAs} + \frac{2GaAs}{Ga + As} + \frac{Ga + As}{2}\right]\left[\frac{Ga}{Ge} + \frac{As}{Ge} + 1\right] \approx CHNOPS\left[\frac{Ga}{Si} + \frac{As}{Si} + 1\right]$$

$$\sqrt{GaAs}\left(\frac{O}{S}\right) + \frac{2GaAs}{Ga + As}\left(\frac{O}{S}\right) + \frac{Ga + As}{2}\left(\frac{O}{S}\right) \approx CHNOPS$$

$$\frac{O}{S} \approx \frac{\left[\frac{Ga}{Ge} + \frac{As}{Ge} + 1\right]}{\left[\frac{Ga}{Si} + \frac{As}{Si} + 1\right]}$$

$$\frac{\sqrt{GaAs}(Ga + As) + 2GaAs}{2(Ga + As)Ge} \approx 1$$

$$\frac{C + H + N + O + P + S}{Ga + As + Ge} \approx \frac{1}{2}$$

We can also make a construct for silicon doped with gallium and phosphorus:

$$(C + N + O + H) \approx \frac{2(Ga + P)Si}{\sqrt{GaP}(Ga + P) + 2GaP}(P + B + Si)$$

$$HNCO \approx \frac{2(Ga + P)Si}{(Ga + P)\left[\sqrt{GaP} + \frac{2GaP}{Ga + P}\right]}(P + B + Si)$$

$$HNCO \approx \frac{2(P + B + Si)Si}{\sqrt{GaP} + \frac{2GaP}{Ga + P}}$$

And we have for germanium doped with gallium and phosphorus:

$$\frac{\sqrt{GaP}(Ga + P) + 2GaP}{2(Ga + P)Ge} \approx \phi$$

$$\left[\sqrt{GaP} + \frac{2GaP}{Ga + P} + \frac{Ga + P}{2}\right]\left[\frac{P}{Ge} + \frac{B}{Ge} + \frac{Si}{Ge}\right] \approx HNCO\left[\frac{Ga}{Ge} + \frac{As}{Ge} + 1\right]$$

$$\sqrt{GaP}\left(\frac{B}{S}\right) + \frac{2GaP}{Ga + P}\left(\frac{B}{S}\right) + \frac{Ga + P}{2}\left(\frac{B}{S}\right) \approx HNCO$$

The Fundamental Albioequations

$$\left[\sqrt{PB} + \frac{2PB}{P+B} + \frac{P+B}{2}\right]\left[\frac{P}{Si} + \frac{B}{Si} + 1\right] \approx HNCO\left[\frac{Ga}{Ge} + \frac{As}{Ge} + 1\right]$$

$$\left[\sqrt{GaAs} + \frac{2GaAs}{Ga+As} + \frac{Ga+As}{2}\right]\left[\frac{Ga}{Ge} + \frac{As}{Ge} + 1\right] \approx CHNOPS\left[\frac{Ga}{Si} + \frac{As}{Si} + 1\right]$$

$$\left[\sqrt{GaP} + \frac{2GaP}{Ga+P} + \frac{Ga+P}{2}\right]\left[\frac{P}{Ge} + \frac{B}{Ge} + \frac{Si}{Ge}\right] \approx HNCO\left[\frac{Ga}{Ge} + \frac{As}{Ge} + 1\right]$$

$$HNCO \approx \frac{2(P+B+Si)Si}{\sqrt{GaP} + \frac{2GaP}{Ga+P}}$$

$$\frac{\sqrt{PB}(P+B) + 2PB}{2(P+B)Si} \approx \phi$$

$$\frac{\sqrt{GaAs}(Ga+As) + 2GaAs}{2(Ga+As)Ge} \approx 1$$

$$\frac{\sqrt{GaP}(Ga+P) + 2GaP}{2(Ga+P)Ge} \approx \phi$$

$$\frac{\sqrt{GaP}(Ga+P) + 2GaP}{2(Ga+P)Si} \approx \Phi$$

$$\frac{C+N+O+H}{P+B+Si} \approx \phi$$

$$\frac{C+H+N+O+P+S}{Ga+As+Ge} \approx \frac{1}{2}$$

$$\frac{Zn}{Se} \approx \frac{\left[\frac{P}{Si} + \frac{B}{Si} + 1\right]}{\left[\frac{Ga}{Ge} + \frac{As}{Ge} + 1\right]}$$

$$\frac{O}{S} \approx \frac{\left[\frac{Ga}{Ge} + \frac{As}{Ge} + 1\right]}{\left[\frac{Ga}{Si} + \frac{As}{Si} + 1\right]}$$

We now want to write out the equations for atomic radius, density, and molar mass as these are the components upon which the properties of the elements should rely.

P_R	Radius Phosphorus	100 pm	
B_R	Radius Boron	85 pm	
Si_R	Radius Silicon	110 pm	
Ga_R	Radius Gallium	130 pm	
As_R	Radius Arsenic	115 pm	
Ge_R	Radius Germanium	125 pm	
P_M	Molar Mas Phosphorus	30.97 g/mol	
B_M	Molar Mass Boron	10.81 g/mol	
Si_M	Molar Mass Silicon	28.09 g/mol	

You will find:

$$\left[\frac{Ga_R}{Ge_R} + \frac{As_R}{Ge_R} + 1\right] \approx \pi \quad \text{and}$$

$$\left[\frac{P_R}{Si_R} + \frac{B_R}{Si_R}\right] \approx \Phi$$

Or,...

$$\frac{P_R + B_R}{Si_R} \approx \Phi \quad \text{Or,....} \quad \frac{Si_R}{P_R + B_R} \approx \phi$$

We now subscript the elements with M for molar mass and find:

$$\frac{\sqrt{P_M B_M}(P_M + B_M) + 2P_M B_M}{2(P_M + B_M)Si_M} \approx \phi$$

Which can be written,...

$$\frac{\sqrt{P_M B_M} + \frac{2P_M B_M}{(P_M + B_M)}}{2Si_M} \approx \phi$$

Which yields,...

$$\frac{2Si_R}{P_R + B_R}Si_M \approx \sqrt{P_M B_M} + \frac{2P_M B_M}{P_M + B_M}$$

AMINO ACIDS

In order to have biological life we need to have carbon. The Astronomer Fred Hoyle figured out how carbon is made by stars. It starts from the formation of helium from hydrogen then from there helium forms carbon in the triple alpha process:

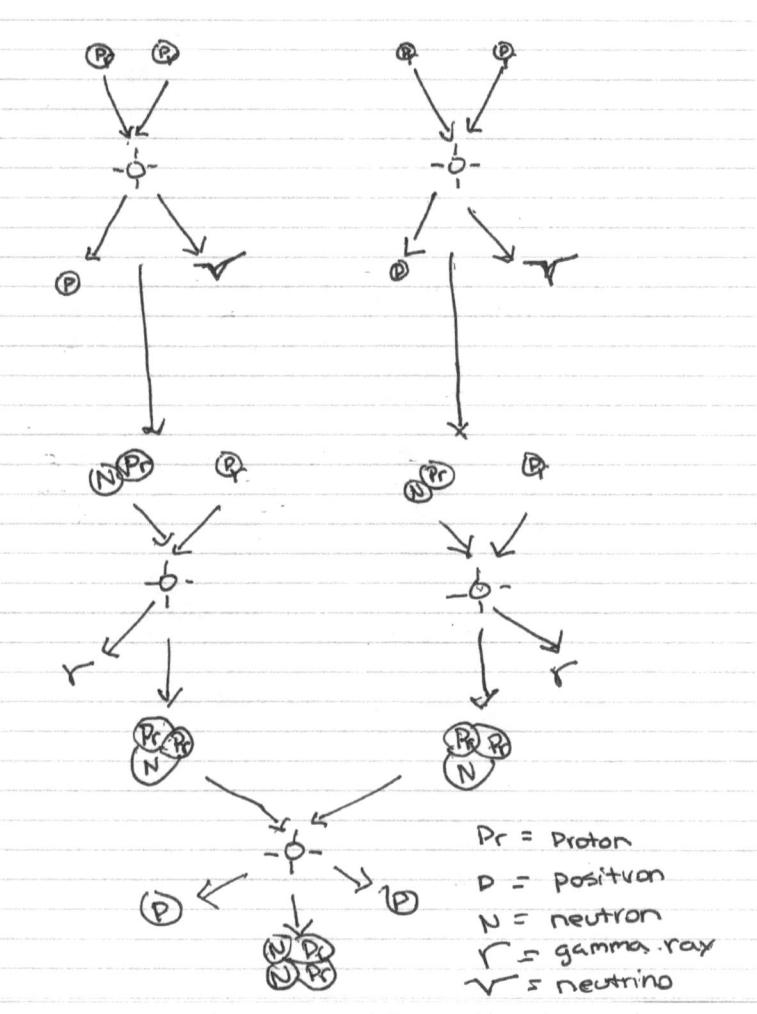

Pr = Proton
D = Positron
N = neutron
Γ = gamma ray
V = neutrino

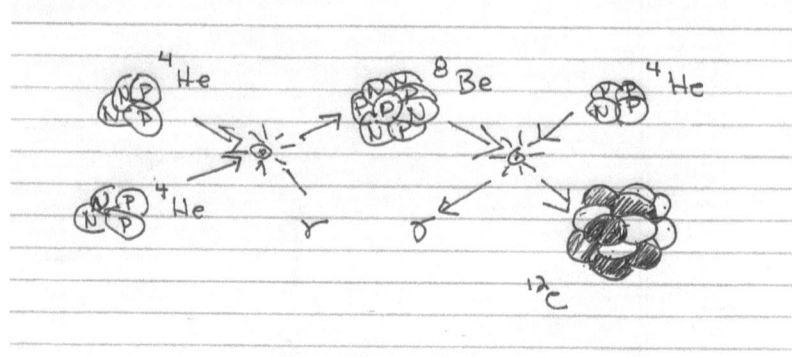

Conducted in 1952 by Stanley Miller and Harold Urey they simulated the primordial conditions thought to exist in the early Earth by mixing Water (H2O) methane (CH4) ammonia (NH3) and hydrogen gas (H2) in a flask and passing a spark through it. The result was the production of 11 of the 20 genetically encoded amino acids. It is thought that the reactions that formed amino acids like this are like a reaction that was recorded a century earlier by Strecker called the strecker synthesis which was:

```
Aldehyde        Ammonia                              Imine    H
    O                                  OH                      /
    ||                                 ||            R--C   + H₂O
R--C--H     +    NH³   ------>    R-C-H   --->        ||
                                     ||              NH         |
                                     NH₂                        | HCN
                                                                |
        H  O                    H  O                   H     V
        |  ||         H₂O       |  ||         H₂O      |
NH₃+ R-C-C--OH  <--------  R-C-C-NH₂  <-----  R-C-C≡N
        |             heat     |                       |
        NH₂                    NH₂                     NH₂
    amino  acid                                   aminocyanonitril
```

I processed the 20 genetically encoded amino acids according to the following scheme:

$$\frac{aminogroup}{acidgroup}(RGroup)$$

In hopes of finding a connection between artificial intelligence and the biological. The result was that two of the amino acids were equal to elements in the periodic table of the elements and they were perfectly carbon (C) the core element of biological life, and silicon (Si) the core element of of artificial intelligence. The amino acids are the building blocks of life, synthesized into proteins by DNA. The two amino acids were serine and glutamine as follows,...

$$\frac{H_3N}{COO}(CH_2 + OH) = C$$

$$\frac{H_3N}{COO}(2CH_2 + CO + NH_2) = Si$$

$$\frac{H_3N}{COO} = (1 - \phi) \qquad \phi = \frac{\sqrt{5} - 1}{2}$$

$$(1 - \phi) = \frac{Si}{Ge}$$

$$\frac{Si}{Ge}(CH_2 + OH) = C$$

$$(CH_2 + OH) = \frac{C}{Si}Ge$$

$$\frac{Si}{Ge}(2CH_2 + CO + NH_2) = Si$$

$$(2CH_2 + CO + NH_2) = Ge$$

Ge is the other core semiconductor element.

$$\frac{(CH_2 + OH)}{(2CH_2 + CO + NH_2)} = \frac{C}{Si}$$

$$C = \frac{(CH_2 + OH)}{(2CH_2 + CO + NH_2)}Si$$

These equation are nearly 100% accurate.

$CH2+OH=12.01+2(1.01)+16,00+1.01=14.03+17.01=31.04$ g/mol

$Si/Ge=28.09/72.61=0.38686$

$(0.38686)(31.04)=12.008$

$C=12.01$

$$\frac{12.008}{12.01}100 = 99.98\%$$

$2CH2=2(12.01+2.02)=28.06$

$CO=12.01+16.00=28.01$

$NH2=14.01+2.02=16.03$

$28.06+28.01+16.03=72.1$

$Ge=72.61$

$$\frac{72.1}{72.61}100 = 99.2976\%$$

The idea is to try to understand biological life, in particular its origins, by looking at something we understand, artificial intelligence.

The equations imply:

$$\frac{H_3N}{COO}Ge \approx Si$$

The primordial compounds from which amino acids are made — water ($H2O$) methane ($CH4$) and ammonia ($NH3$) —seem to be related to primitive AI which would be a tungsten filament (W) encased in a glass tube (SiO2) to make vacuum tubes for switches as follows:

$$\frac{W}{SiO_2} \approx \frac{H_2O}{CH_4} + \frac{NH_3}{CH_4} + 1$$

DNA

In order to have DNA and RNA we need the sugars deoxyribose and ribose. But sugars are polymers of CH2O which itself is not a sugar but has the same structure as a sugar which is (CH2O)n. When we consider the Miller-Urey experiment where Miller and Urey produced from the hypothesized primordial earth composition of NH3 (ammonia), CH4 (methane) and H2O (water) some of the 20 genetically encoded amino acids that are the building blocks of life, then it is interesting that these are equal to the ancient, "primordial" artificial intelligence which were switches made of tungsten filaments (W) encased in glass (SiO2):

$$\frac{W}{SiO_2} \approx \left[\frac{H_2O}{CH_4} + \frac{NH_3}{CH_4} + 1\right]$$

Because the monomer from which sugars were surely made as DNA came into existence on the primordial earth has it equivalence in

$$\frac{SiO_2}{CH_2O} = 2$$

$$\frac{\left[\frac{Ga}{Si} + \frac{As}{Si} + 1\right]}{\left[\frac{Ga}{Ge} + \frac{As}{Ge} + 1\right]} \approx 2$$

So that,...

$$CH_2O \approx SiO_2 \frac{\left[\frac{Ga}{Si} + \frac{As}{Si} + 1\right]}{\left[\frac{Ga}{Ge} + \frac{As}{Ge} + 1\right]}$$

Where Ga and Ge are gallium and arsenic, doping agents for germanium (Ge) and silicon (Si) the first and second generation semiconductors, respectively.

And indeed DNA consists of the sugar deoxyribose, but it is made as well of the nitrogenous bases adenine, guanine, cytosine, and thymine. These are:

Adenine=C5H5N5=135.13 g/mol

Guanine=C5H5N5O=151.13 g/mol

Cytosine=C5H4N3O=111.1 g/mol

Thymine=C5H6N2O2=126.113 g/mol

And we have the sum of their ratios is pi, the ratio of the circumference of a circle to its diameter:

$$\frac{C_5H_5N_5}{C_5H_6N_2O_2} + \frac{C_5H_5N_5O}{C_5H_6N_2O_2} + \frac{C_5H_4N_3O}{C_5H_6N_2O_2} \approx \pi$$

And that,...

$$\frac{C_5H_5N_5}{C_5H_6N_2O_2} + \frac{C_5H_5N_5O}{C_5H_6N_2O_2} + \frac{C_5H_4N_3O}{C_5H_6N_2O_2} \approx \frac{Ga}{Ge} + \frac{As}{Ge} + 1$$

And we have to consider its phosphate backbone, H3PO4

$$\frac{Se}{Zn} \approx \frac{H_3PO_4}{C_5H_6N_2O_2}$$

Where ZnSe is zinc selenide, an intrinsic semiconductor.

DNA is, as well, approximately equal to the primordial prebiotic substances:

$$\frac{C_5H_5N_5}{C_5H_6N_2O_2} + \frac{C_5H_5N_5O}{C_5H_6N_2O_2} + \frac{C_5H_4N_3O}{C_5H_6N_2O_2} \approx \frac{H_2O}{CH_4} + \frac{NH_3}{CH_4} + 1$$

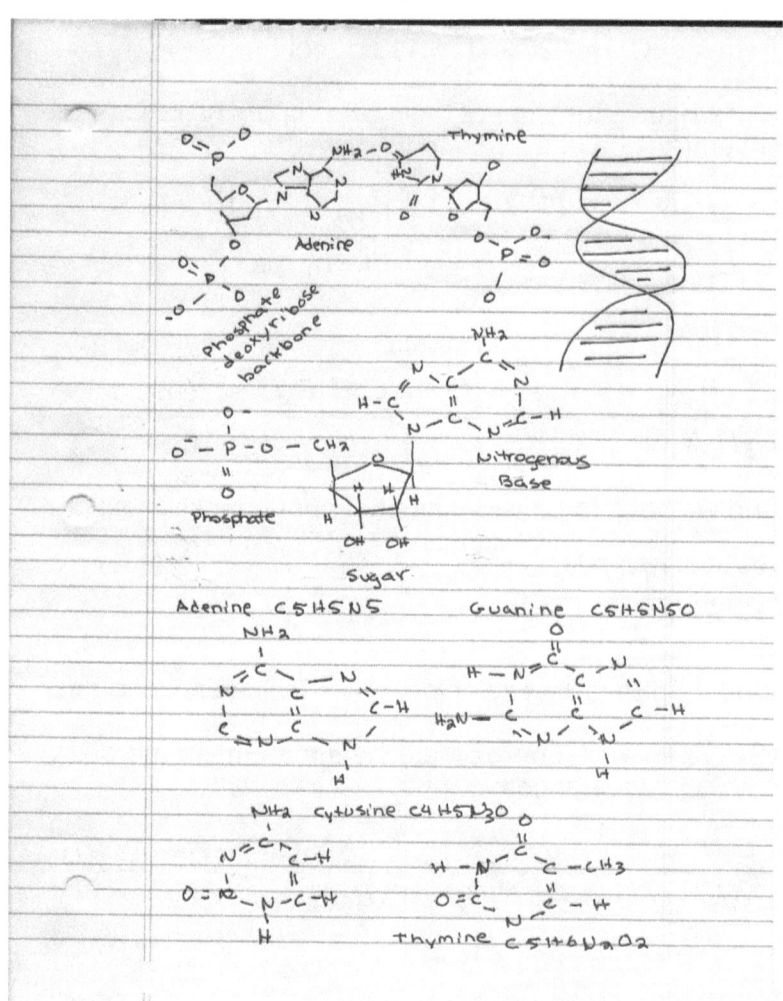

MALE AND FEMALE IN AI

We consider the female sex hormone estradiol (estrogen , E):

$$C_{18}H_{24}O_2 = 272.38 g/mol$$

And the male sex hormone testosterone (T):

$$C_{19}H_{28}O_2 = 288.42 g/mol$$

And, cholesterol (Ch) from which both are made:

$$C_{27}H_{46}O = 386.65 g/mol$$

And notice,...

$$\frac{Ch + T}{E} = 2.5$$

And we consider the semiconductor materials used to make AI:

$$\frac{Ge}{Si} = 2.6$$

And write,...

$$\frac{Ch + T}{E} = \frac{Ge}{Si}$$

$$T = \frac{Ge}{Si}E - Ch \qquad E = \frac{Si}{Ge}(T + Ch)$$

$$T\left(1 - \frac{Si}{Ge}\right) + E\left(1 - \frac{Ge}{Si}\right) = Ch\left(\frac{Si}{Ge} - 1\right)$$

We notice that the masculine (T) is in inverse relation to the feminine (E), but that the two add up to on whole (Ch) in that the masculine has coefficient 1-Si/Ge and the feminine has coefficient 1-Ge/Si. This expresses the inverse relationships between man and woman.

ASYMMETRY IN AI

While artificial intelligence is characterized by Stokes theorem, biological life is characterized by the quadratic formula. The connection between the two is through the golden ratio. We begin with artificial intelligence.

ARTIFICIAL INTELLIGENCE

The primary elements of artificial intelligence (AI) used to make diodes and transistors, silicon (Si) and germanium (Ge) doped with boron (B) and phosphorus (P) or gallium (Ga) and arsenic (As) have an asymmetry due to boron. Silicon and germanium are in group 14 like carbon (C) and as such have 4 valence electrons. Thus to have positive type silicon and germanium, they need doping agents from group 13 (three valence electrons) like boron and gallium, and to have negative type silicon and germanium they need doping agents from group 15 like phosphorus and arsenic. But where gallium and arsenic are in the same period as germanium, boron is in a different period than silicon (period 2) while phosphorus is not (group 3). Thus aluminum (Al) is in boron's place. This results in an interesting equation.

$$\frac{Si(As - Ga) + Ge(P - Al)}{SiGe} = \frac{2B}{Ge + Si}$$

The differential across germanium crossed with silicon plus the differential across silicon crossed with germanium normalized by the product between silicon and germanium is equal to the boron divided by the average between the germanium and the silicon. The equation has nearly 100% accuracy:

$$\frac{28.09(74.92 - 69.72) + 72.61(30.97 - 26.98)}{(28.09)(72.61)} = \frac{2(10.81)}{(72.61 + 28.09)}$$

$$0.213658912 = 0.21469712$$

$$\frac{0.213658912}{0.21469712} = 0.995$$

$$99.5\%$$

To illustrate this asymmetry here is that section of the periodic table:

	13	14	15
Period 2	B		
Period 3		Si	P
Period 4	Ga	Ge	As

This I believe will open up the door to a lot in the periodic table of the elements because:

$$\left| \begin{pmatrix} 0 & 0 & Si \\ As & Ga & 0 \\ 1 & 1 & 0 \end{pmatrix} \right| = Si(As - Ga)$$

$$\left| \begin{pmatrix} 0 & 0 & Ge \\ P & Al & 0 \\ 1 & 1 & 0 \end{pmatrix} \right| = Ge(P - Al)$$

But,..

$$\frac{2B}{Ge + Si} = \frac{B}{\frac{Ge + Si}{2}} \quad \text{And,}$$

$$\frac{Ge + Si}{2} = average(Ge, Si)$$

But,

$$average(f) = \frac{1}{b - a} \int_a^b f(x) dx$$

And (Stokes theorem)

$$\int_S (\nabla \times \vec{u}) \cdot d\vec{S} = \oint_C \vec{u} \cdot d\vec{r}$$

Where,

$$\nabla \times \vec{u} = \left\| \begin{pmatrix} \vec{i} & \vec{j} & \vec{k} \\ \frac{\partial}{\partial x} & \frac{\partial}{\partial y} & \frac{\partial}{\partial z} \\ u_1 & u_2 & u_3 \end{pmatrix} \right\|$$

But

$$\frac{Si(As - Ga) + Ge(P - Al)}{SiGe} = \frac{2B}{Ge + Si}$$

Can be written

$$\frac{Si}{B}(As - Ga) + \frac{Ge}{B}(P - Al) = \frac{2SiGe}{Si + Ge}$$

But

$$\frac{2SiGe}{Si + Ge}$$

Is the harmonic mean between Si and Ge. However, where we had

$$average(f) = \frac{1}{b - a} \int_a^b f(x)dx$$

We have

$$harmonic(f) = \frac{1}{\frac{1}{b-a} \int_a^b f(x)^{-1}dx}$$

As it would turn out

$$\frac{Si}{B}(As - Ga) = \frac{1}{3}H$$

$$\frac{Ge}{B}(P - Al) = \frac{2}{3}H$$

Where

$$H = \frac{2SiGe}{Si + Ge}$$

We have

$$\bar{f} = \frac{1}{b - a}\int_{a}^{b} f(x)dx$$

Since we want f(x) such that

$$\frac{1}{Ge - Si}\int_{Si}^{Ge} f(x)dx = \frac{Si + Ge}{2}$$

Then,

$$f(x) = x$$

But,

$$harmonic(f) = \frac{1}{\frac{1}{b-a}\int_{a}^{b} f(x)^{-1}dx}$$

$$\frac{Si}{B}(As - Ga) + \frac{Ge}{B}(P - Al) = \frac{Ge - Si}{\int_{Si}^{Ge} \frac{dx}{x}}$$

But,

$$\frac{2SiGe}{Si + Ge} \approx \frac{Si + Ge}{2}$$

$$\int_0^1 \int_0^1 \left[\frac{Si}{B}(As - Ga) + \frac{Ge}{B}(P - Al)\right] dxdy \approx \frac{1}{Ge - Si}\int_{Si}^{Ge} xdx$$

$$(Ge - Si)\left[\frac{Si}{B}(As - Ga) + \frac{Ge}{B}(P - Al)\right] \approx \int_{Si}^{Ge} xdx$$

$$(72.61 - 28.09)\left[\frac{28.09}{10.81}(74.92 - 69.72) + \frac{69.72}{10.81}(30.97 - 26.98)\right] = \frac{1}{2}(72.61^2 - 28.09^2)$$

$$1746.5 \approx 2241.5$$

$$80\%$$

I propose the following two matrices:

$$\left\|\begin{pmatrix} \vec{i} & \vec{j} & \vec{k} \\ \frac{\partial}{\partial x} & \frac{\partial}{\partial y} & \frac{\partial}{\partial z} \\ 0 & \frac{Si}{B}(Ga)z & \frac{Si}{B}(As)y \end{pmatrix}\right\| = \frac{Si}{B}(As - Ga)\vec{i}$$

$$\left|\left|\begin{pmatrix} \vec{i} & \vec{j} & \vec{k} \\ \frac{\partial}{\partial x} & \frac{\partial}{\partial y} & \frac{\partial}{\partial z} \\ \frac{Ge}{B}(Al)z & 0 & \frac{Ge}{B}(P)x \end{pmatrix}\right|\right| = \frac{Ge}{B}(P - Al)\vec{j}$$

And we have said:

$$\frac{2SiGe}{Si + Ge} \approx \frac{Si + Ge}{2}$$

And,

$$\frac{Si}{B}(As - Ga) = \frac{1}{3}H$$

$$\frac{Ge}{B}(P - Al) = \frac{2}{3}H$$

Where

$$H = \frac{2SiGe}{Si + Ge}$$

Thus if

$$\nabla \times \vec{u} = \frac{Si}{B}(As - Ga)\vec{i}$$

$$\nabla \times \vec{u} = \frac{Ge}{B}(P - Al)\vec{j}$$

$$\vec{dS} = dxdy$$

Then we have the following two integrals:

$$\int_0^1 \int_0^1 \frac{Si}{B}(As - Ga)dxdy = \frac{1}{3}\frac{1}{Ge - Si}\int_{Si}^{Ge} xdx$$

$$\int_0^1 \int_0^1 \frac{Ge}{B}(P - Al)dxdy = \frac{2}{3}\frac{1}{Ge - Si}\int_{Si}^{Ge} xdx$$

When we say we want f(x) such that

$$\frac{1}{Ge - Si}\int_{Si}^{Ge} f(x)dx = \frac{Si + Ge}{2}$$

It is true but the equations are only 80% accurate so we want f(x) such that

2241.5x=1746.5 or x=0.8=8/10 so that our equations become

$$\int_0^1 \int_0^1 \left[\frac{Si}{B}(As - Ga) + \frac{Ge}{B}(P - Al)\right]dxdy \approx \frac{1}{Ge - Si}\frac{4}{5}\int_{Si}^{Ge} xdx$$

And

$$\int_0^1 \int_0^1 \frac{Si}{B}(As - Ga)dxdy = \frac{1}{4}\frac{1}{Ge - Si}\int_{Si}^{Ge} xdx$$

$$\int_0^1 \int_0^1 \frac{Ge}{B}(P - Al)dxdy = \frac{1}{2}\frac{1}{Ge - Si}\int_{Si}^{Ge} xdx$$

That is for the first equation

$$f(x) = \frac{4}{5}x$$

This gives

$$H\bar{f} = \frac{1}{\frac{1}{72.61 - 28.09}\int_{28.09}^{72.61}\frac{5}{4}\frac{dx}{x}} = \frac{1}{0.002246\frac{5}{4}ln\frac{Ge}{Si}} = 37.6$$

Where $H\bar{f}$ means the harmonic mean of f. Does this approximately equal

$$\frac{2SiGe}{Si + Ge}?$$

$$\frac{2(28.09)(72.61)}{28.09 + 72.61} = 40.5$$

It does.

BIOLOGICAL LIFE

Now for biological life. Life can be considered as based on the physical water H20

$$H_2O = 18g/mol$$

Air, which is about 75% N2 and 25% O2

$$air = 0.75N_2 + 0.25O_2 = 29g/mol$$

And the organic of which the most basic organic compound is isocyanic acid (HNCO)

$$HNCO = 43g/mol$$

We notice

$$\frac{H_2O}{air} \approx \phi$$

Where ϕ (phi) is the golden ratio conjugate. The golden ratio and the golden ratio conjugate are the solution of the quadratic

$$\left(\frac{a}{b}\right)^2 - \frac{a}{b} - 1 = 0 \text{ that meets the conditions}$$

$$\frac{a}{b} = \frac{b}{c} \text{ and } a=b+c$$

Where $\Phi = \dfrac{a}{b}$ and $\phi = \dfrac{\sqrt{5}-1}{2}$, $\phi = \dfrac{1}{\Phi}$.

We also notice

$$air - H_2O \approx C$$

We have

$$\frac{air - C}{air} \approx \phi$$

$$\frac{H_2O}{H_2O + C} \approx \phi$$

We also notice

$$H(air - H_2O) \approx C$$

Thus,...

$$\frac{H_2O}{H_2O + C} \approx \frac{air - C}{air}$$

$$\frac{C^2}{air} + C\frac{H_2O}{air} - C \approx 0$$

And finally

$$\frac{C^2}{air} + C\frac{H_2O}{air} - (air - H_2O) \approx 0$$

For the quadratic

$$ax^2 + bx + c = 0$$

The solution is

$$x = \frac{-b \pm \sqrt{b^2 - 4ac}}{2a}$$

Then,...

$$C = \frac{1}{2}air\left(-\frac{H_2O}{air} \pm \sqrt{\frac{(H_2O)^2}{(air)^2} - \frac{4(H_2O - air)}{air}}\right)$$

C=10.991 g/mol

$$\frac{10.991}{12.01} = 0.915$$

91.5% accurate

DIFFERENTIAL EQUATIONS

If we start with carbon (C) in group 14 period 2 as one moving across the period with value 12 g/mol to primary biological element nitrogen (N) with value 14 g/mol then to oxygen (O) with value 16 g/mol it is quite easy to see this can be fit with

$$y = 2x + 10$$

And, if we move down from core biological element carbon down the group 14 to core Al element silicon (Si) with value 28 g/mol then to core Al element germanium (Ge) with value 72.61 g/mol it is more of an effort, but we can fit this curve, it is

$$y = \frac{7}{2}e^x + 4$$

We see the first is the solution of the differential equation

$$y\frac{dy}{dx} - 4x - 20 = 0$$

And the second is the solution to the differential equation

$$\frac{dy}{dx} - y + 4 = 0$$

We see this in the first instant because

$$\frac{1}{2}y\frac{dy}{dx} - 2x - 10 = 0$$

And if the second by introducing the integration factor

$$\mu(x) = e^{-\int dx} = e^{-x}$$

$$\frac{dy}{dx}e^{-x} - e^{-x}y = -4e^{-x}$$

Which is by the product rule

$$\frac{d}{dx}\left(e^{-x}y\right) = -4e^{-x}$$

Integrating

$$e^{-x}y = 4e^{-x} + C$$

Or,... $y = 4 + Ce^x$

If C=7/2, then $y = \frac{7}{2}e^x + 4$

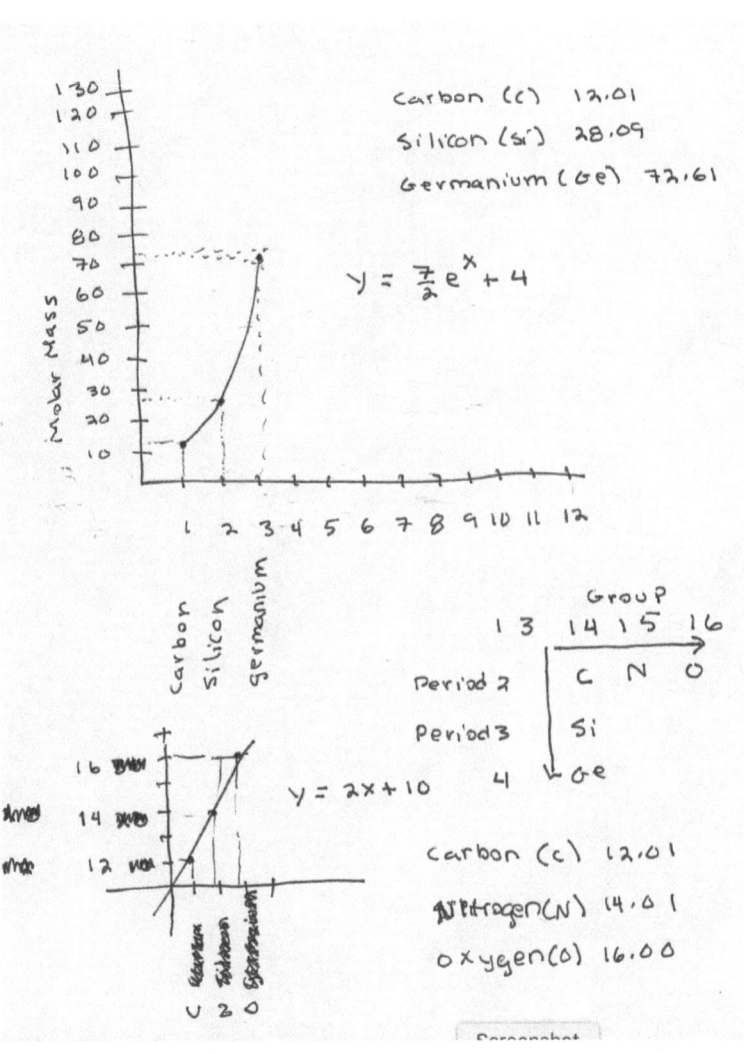

Carbon (C)   12.01
Silicon (Si)   28.09
Germanium (Ge)   72.61

$$y = \frac{7}{2}e^x + 4$$

Molar Mass

130
120
110
100
90
80
70
60
50
40
30
20
10

1 2 3 4 5 6 7 8 9 10 11 12

carbon
silicon
germanium

Group

            13   14 15   16

Period 2   C   N   O

Period 3   Si

            4   Ge

carbon (C) 12.01
Nitrogen (N) 14.01
oxygen (O) 16.00

16

14

12

$$y = 2x + 10$$

2 0

CROSS PRODUCT

We have the following scenario

$$
\begin{array}{cccc}
 & 14 & 15 & 16 \\
\text{Period 2.} & C & N & O \\
\text{Period 3} & Si & & \\
\text{Period 4.} & Ge & &
\end{array}
$$

So we want to cross C, N, O with C, Si, Ge

$$
\begin{pmatrix} \vec{i} & \vec{j} & \vec{k} \\ C & N & O \\ C & Si & Ge \end{pmatrix} = \begin{pmatrix} \vec{i} & \vec{j} & \vec{k} \\ 12.01 & 14.01 & 16.00 \\ 12.01 & 28.09 & 72.61 \end{pmatrix} =
$$

$$
567.8261\,\vec{i} - 679.8861\,\vec{j} + 169.1008\,\vec{k} =
$$

$$
\sqrt{322426.47 + 4672245.109 + 28595.08056} =
$$

$$
901.8129902 g/mol
$$

This number looks very familiar to me. In my earlier work comparing biological life to artificial intelligence including amino acids, DNA, elements and their compounds, the most interesting to me was the case of bone. So much so that I wrote a book on it called bone.

The mineral component of bone is:

$$
Ca_5(PO_4)_3OH = 502.32 g/mol
$$

And the organic component is

$$
C_{57}H_{91}N_{19}O_{16} = 1298.67 g/mol
$$

Let us take the arithmetic mean between these two:

$$\frac{502.32 + 1298.67}{2} = 900.495$$

$$\frac{900.495}{901.8129902} = 0.9985$$

That is 99.85% accuracy. In other words

$$\sqrt{(N * Ge - O * Si)^2 + (C * Ge - O * C)^2 + (C * Si - N * C)^2}$$

$$\approx \frac{Ca_5(PO_4)_3OH + C_{57}H_{91}N_{19}O_{16}}{2}$$

INTEGRATING

Returning to our equations for molar mass as a function of elemental number moving across the periods and down the groups starting with carbon in both cases, we have

$$y = 2x + 10$$

And,

$$y = \frac{7}{2}e^x + 4$$

Which are solutions to the differential equations

$$y\frac{dy}{dx} - 4x - 20 = 0$$

And,

$$\frac{dy}{dx} - y + 4 = 0$$

Respectively, and we can integrate them:

$$F(b) - F(a) = \int_1^3 (2x + 10)dx =$$

$$x^2 + 10x \left\{ \begin{matrix} 3 \\ 1 \end{matrix} \right. = (9 + 30) - (1 + 10) = 39 - 11 = 28$$

$$F(b) - F(a) = \int_1^3 \frac{7}{2}e^x dx + \int_1^3 4dx =$$

(70.299+12)-(9.51+4)=82.299-13.51=68.789

$$\frac{68.789}{28} = 2.5 \approx \frac{Ge}{Si} \approx \frac{Ga}{Al} \approx \frac{As}{P}$$

$$f(x) = 2x + 10$$
$$g(x) = \frac{7}{2}e^x + 4$$

$$f(1) = C, f(2) = N, f(3) = O$$
$$g(1) = C, g(2) = Si, g(3) = Ge$$

$$\int_1^3 \frac{7}{2}e^x dx + \int_1^3 4dx = \left(\frac{Ge}{Si}, \frac{Ga}{Al}, \frac{As}{P}\right) \cdot \int_1^3 (2x + 10)dx$$

The Author

www.ingramcontent.com/pod-product-compliance
Lightning Source LLC
Chambersburg PA
CBHW030745200526
45160CB00010B/57/J